MOSQUITOES
by Liza Jacobs

BLACKBIRCH PRESS

San Diego • Detroit • New York • San Francisco • Cleveland • New Haven, Conn. • Waterville, Maine • London • Munich

© 2003 by Blackbirch Press™. Blackbirch Press™ is an imprint of The Gale Group, Inc., a division of Thomson Learning, Inc.

Blackbirch Press™ and Thomson Learning™ are trademarks used herein under license.

For more information, contact
The Gale Group, Inc.
27500 Drake Rd.
Farmington Hills, MI 48331-3535
Or you can visit our Internet site at http://www.gale.com

ALL RIGHTS RESERVED
No part of this work covered by the copyright hereon may be reproduced or used in any form or by any means—graphic, electronic, or mechanical, including photocopying, recording, taping, Web distribution or information storage retrieval systems—without the written permission of the copyright owner.

Every effort has been made to trace the owners of copyrighted material.

Photographs © 1998 by Pan Jian-Hong

Cover Photograph © Corel

© 1998 by Chin-Chin Publications Ltd.

No. 274-1, Sec.1 Ho-Ping E. Rd., Taipei, Taiwan, R.O.C.
Tel: 886-2-2363-3486 Fax: 886-2-2363-6081

LIBRARY OF CONGRESS CATALOGING-IN-PUBLICATION DATA

Jacobs, Liza.
  Mosquitoes / Liza Jacobs.
    v. cm. — (Wild wild world)
  Includes bibliographical references (p. 24).
  Contents: Blood suckers -- Water lovers -- Mating -- Many kinds.
  ISBN 1-4103-0046-3 (hardback : alk. paper)
  1. Mosquitoes--Juvenile literature. [1. Mosquitoes.] I. Title. II. Series.

QL536.J23 2003
595.77'2--dc21   2003001489

Printed in Taiwan
10 9 8 7 6 5 4 3 2 1

Table of Contents

Blood Suckers 4

Insect Bodies 6

Water Lovers 8

Mating 10

Laying Eggs 12

Four Stages of Growth 14

Larvae 16

Pupae 18

Adults Break Free 20

For More Information 24

Glossary 24

Blood Suckers

Mosquitoes have lived on the earth for millions of years. These insects live all over the world. Mosquitoes drink the nectar from flowers and other plants. Female mosquitoes also drink the blood of animals and humans. The blood has the nutrients a female needs for her eggs. Mosquitoes suck their food through a long tube called a proboscis.

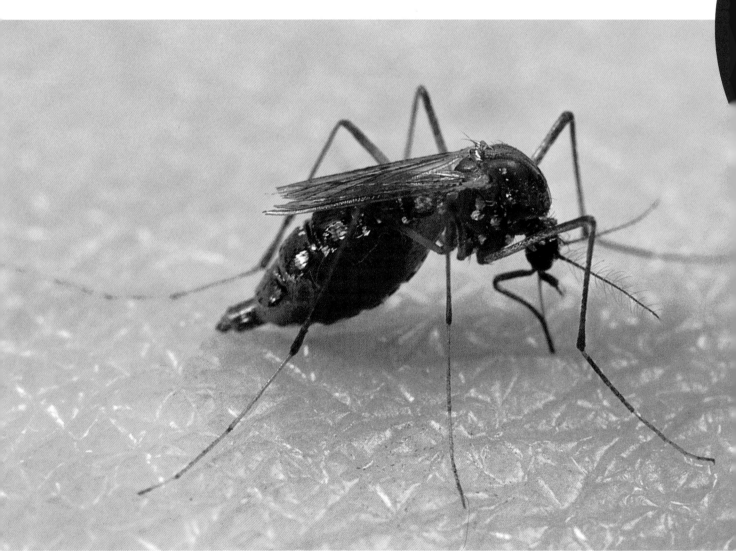

A female with a full abdomen of blood finishes her meal on the surface of a human's skin.

Insect Bodies

Like all insects, the body of a mosquito has three main parts. It has a head, a thorax (midsection), and an abdomen (rear section). The thorax is behind the head. A mosquito's wings and six long legs are attached to the thorax. The abdomen is behind the thorax. Mosquitoes have large compound eyes. Compound eyes are made up of hundreds of tiny eyes. Male mosquitoes have long feelers, or palps. They also have large, feathery antennae. Females have shorter palps and much thinner antennae.

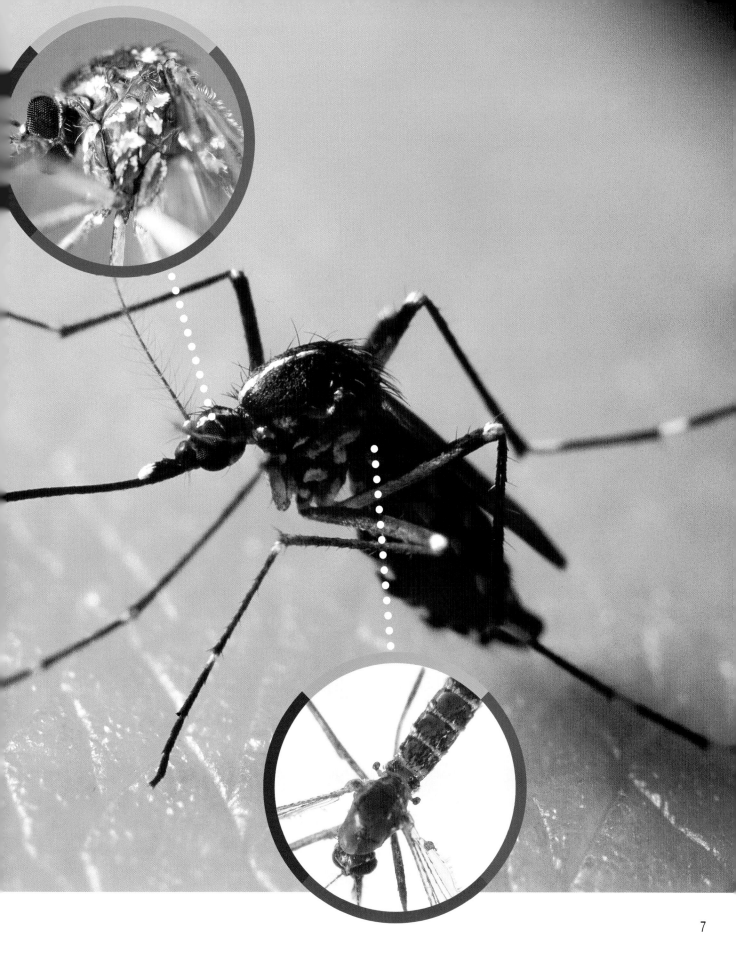

Water Lovers

Mosquitoes live in warm, damp parts of forests, grasslands, mountains, and swamps. They love water and are found near lakes, ponds, and rivers. In fact, they live everywhere on earth except in deserts or polar areas. Mosquitoes also use water for laying their eggs.

Mating

Mosquitoes only live for about 2 weeks. In that time, females may lay eggs 4 or 5 times! When a female mosquito is ready to mate, it makes a high-pitched buzzing sound. This attracts male mosquitoes to her. Even in the dark, males can find a female that is making this sound.

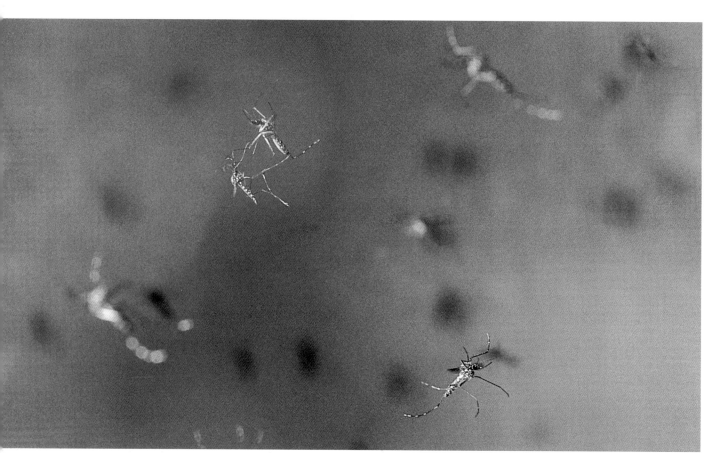

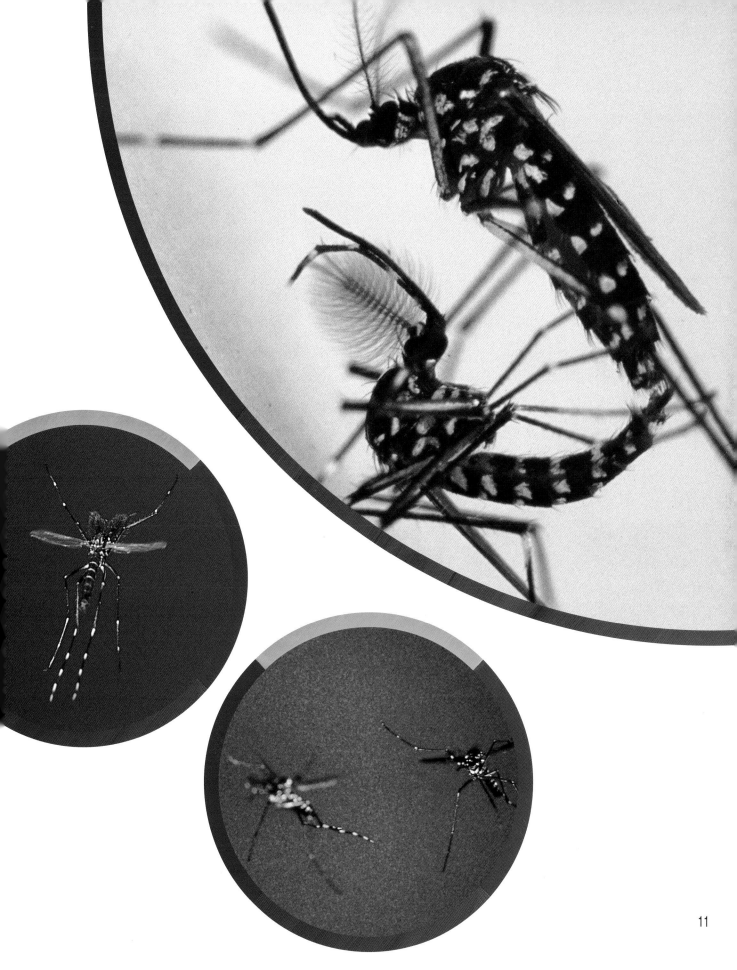

Laying Eggs

After mating, some female mosquitoes lay their eggs in the water. To do this, the female puts the end of her abdomen in the water. Others lay their eggs on the surface of the water. The source of water does not much matter. It can be a clean body of water or just a large puddle of dirty water. Different kinds of mosquitoes lay different kinds of eggs. Some lay many separate eggs. Others lay eggs grouped together in a cluster called a raft. A female can lay hundreds of eggs at a time.

Four Stages of Growth

Like many insects, mosquitoes go through four stages of life as they grow into their adult forms. The four stages are egg, larva, pupa, and adult. The first three stages of a mosquito's life are lived in the water. Mosquito eggs hatch into larvae within a few days. Also called wrigglers, these little larvae do not have any legs.

Larvae

Larvae look like tiny worms. They swim around in the water swallowing small particles of food. Larvae must come to the surface to breathe. They have a tube-like body part called a siphon. Wrigglers hang upside-down in the water with their siphon sticking above the surface. Air travels through the tube, letting the larvae breathe surface air while underwater.

Fish love to eat mosquito larvae. The larvae that do not become fish food grow for 5 to 20 days. As they grow, they molt, or shed their skin, about 4 times. Then the larvae are ready to become pupae.

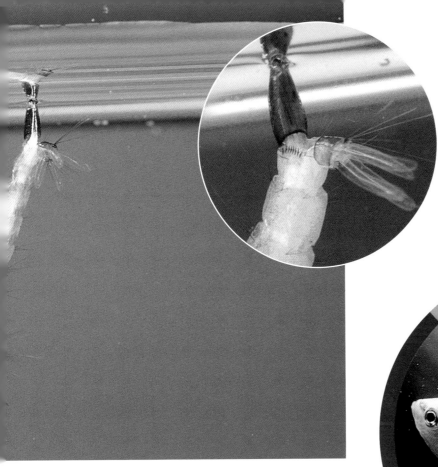

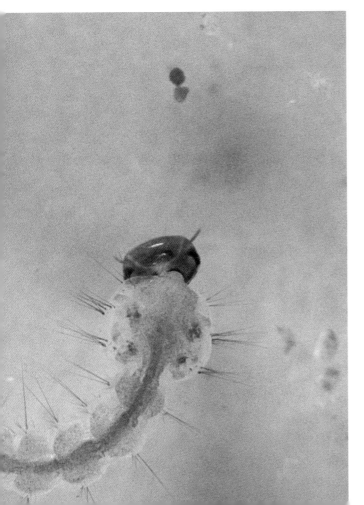

Pupae

This third stage of life lasts between 2 and 8 days. During this time, a pupa does not eat. It begins to form the body parts of an adult. A pupa's body is curled in the shape of a comma. Pupae are also called tumblers because of the way their comma-shaped bodies swim through the water. As a pupa quickly curls its tail out and back it looks as though it's tumbling around. Like larvae, pupae also go to the surface to breathe. They use a pair of breathing tubes called trumpets.

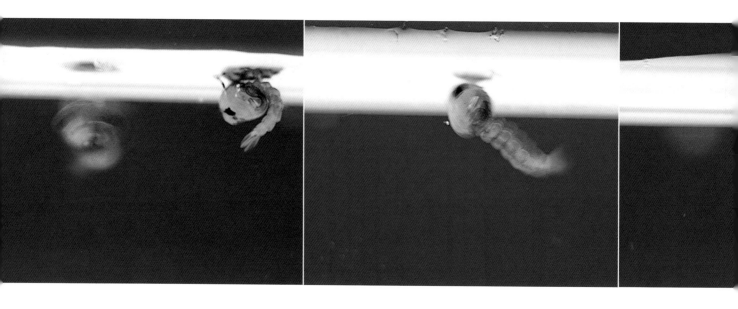

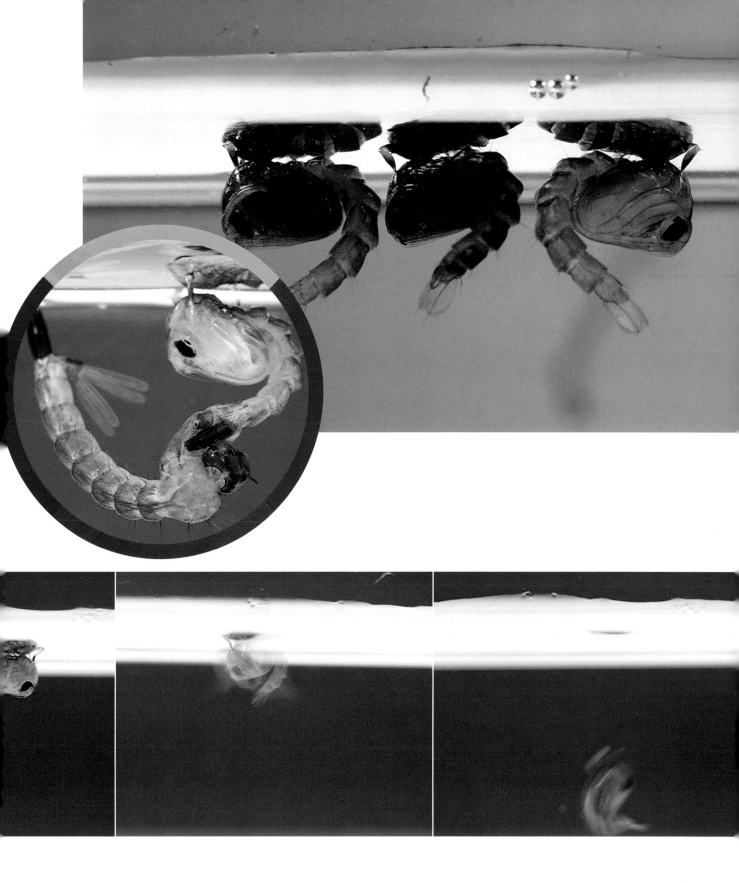

Adults Break Free

It only takes a few days for a pupa to turn into an adult mosquito. When it is ready, the pupa skin splits open and the adult breaks free. The soft, wet insect comes out of the water to dry.

Soon, its outer covering hardens and its wings dry. It is ready to fly off in search of a mate. The pupa skin is all that is left behind in the water.

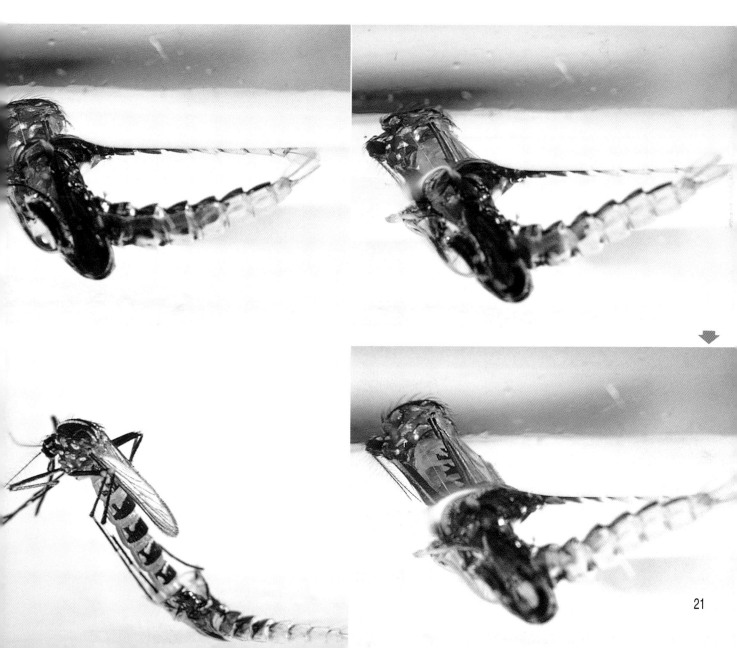

There are more than 2,000 different kinds of mosquitoes. Most mosquitoes cannot hurt people. But there are some types that carry diseases. Disease is sometimes passed to a person who is bitten through a mosquito's saliva.

As much as mosquitoes can bother humans, they are an important food source for many animals. Birds, fish, frogs, beetles, bats, and many other kinds of animals feed on mosquitoes. As part of the food chain, these creatures play an important role in the balance of nature.

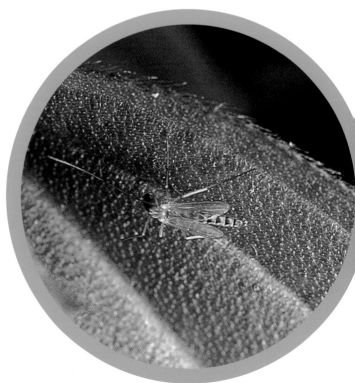

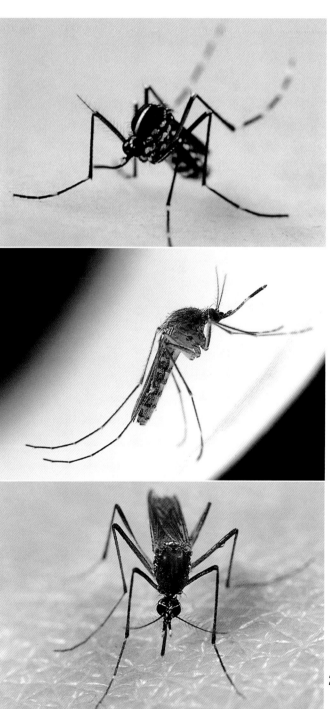

For More Information

Goor, Ron and Nancy. *Insect Metamorphosis: From Egg to Adult.* New York: Atheneum, 1990.

Meister, Cari. *Mosquitoes.* Edina, MN: ABDO Publishing, 2001.

Glossary

compound eye an eye that has many lenses

molt to shed the outer skin or covering

proboscis the long tube a mosquito uses to suck up its food

raft a cluster of mosquito eggs